The Old English Game Fowl
The History, Management, Breeding and Feeding of Game Fowls

by Herbert Atkinson

with an introduction by Jackson Chambers

This work contains material that was originally published in 1891.

This publication is within the Public Domain.

*This edition is reprinted for educational purposes
and in accordance with all applicable Federal Laws.*

Introduction Copyright 2018 by Jackson Chambers

Self Reliance Books

Get more historic titles on animal and stock breeding, gardening and old fashioned skills by visiting us at:

http://selfreliancebooks.blogspot.com/

Disclaimer

This book was written in an age when cock-fighting was widely acceptable throughout society. In many places throughout the world, cock-fighting has been made illegal.

The material presented herein is intended to be strictly for educational purposes with the purpose of enlightening Game Fowl breeders about the history of their breed. Publication of the material is neither an endorsement, nor a criticism of its contents. This book is presented as part of large series of educational material on the history and raising of numerous chicken breeds for utility or exhibition purposes.

As the reader, please consider it your duty to become familiar with local, state, provincial and federal laws relating to the subject matter contained herein before attempting to utilize any of the information presented.

As the author, publisher and retailer cannot control how the reader utilizes the historical information presented in the pages herein, they hereby disclaim any liability to any party for any loss, damage, disruption or other liability that may be incurred by the reader's misuse of this material.

Introduction

I am pleased to present this third title in the "Game Fowl" series.

The work is in the Public Domain and is re-printed here in accordance with Federal Laws.

Though this work is a century old it contains much information on poultry that is still pertinent today.

As with all reprinted books of this age that are intended to perfectly reproduce the original edition, considerable pains and effort had to be undertaken to correct fading and sometimes outright damage to existing proofs of this title. At times, this task is quite monumental, requiring an almost total "rebuilding" of some pages from digital proofs of multiple copies. Despite this, imperfections still sometimes exist in the final proof and may detract from the visual appearance of the text.

I hope you enjoy reading this book as much as I enjoyed making it available to readers again.

Jackson Chambers

BLACK BREASTED LIGHT RED COCK WITH PARTRIDGE AND WHEATEN HENS.

INDEX.

	PAGE
Introduction	7
Origin and History	9
The Revival of Old English Game in the Show Pen	16
The Points of Game Fowls	19
The Different Breeds and Colours	26
The Sport of Cocking	43
The Useful Properties of Old English Game	55
Management and Mating of the Breeding Stock	57
Sitting, Hatching, Rearing, &c.	60
Exhibiting	63
Diseases	64

INTRODUCTION.

In writing this little work on "Old English Game," it has been my endeavour to describe the birds for the benefit of those fanciers and beginners who are not yet fully conversant with them.

I am not for one moment supposing that any work such as the present will be of much value to "Old Hands," but if it causes the true Game Fowl to be better known and more widely kept, and is of any assistance to the poultry fancier about to take up the breed, it will have attained its object.

I have received some of the most interesting information on "Hennies," &c., from a gentleman well known under his *nom de plume* of "Game Cock," whose kindness I take this opportunity to acknowledge. I may add that had I not been frequently asked to write something of this kind by several fanciers, this monograph would probably never have appeared.

March 25th, 1891.   H. A.

THE OLD ENGLISH GAME FOWL.

CHAPTER I.

ORIGIN AND HISTORY.

AT what time the game fowl was first kept in a state of domestication is not known. As in the case of the horse and the dog, the past is silent on that point, and we are left to conjecture. As to the origin of the game cock, many writers have aired their pet theories; nearly all, however, attribute his origin to four varieties of jungle fowl, viz., Gallus ferrugineus, Gallus Sonneratii, Gallus furcatus, and the Ceylon jungle fowl. We are bound, however, to add that, although these wild varieties are game in the wild state, yet the crosses that have been tried in India with the jungle fowl have been failures in point of courage, and while it was the opinion of one of our ablest authorities on game fowls, and in which I entirely coincide, that the yellow-legged and yellow-fleshed varieties were introduced from India at a comparatively later date, and are descended from some of the wild jungle fowls still in existence there, the older white-legged and white-fleshed birds are not of that origin, but have been in England from the earliest times, and are probably descended from some species now extinct. The opinion expressed by Trevor Dickins, Esq., was that the original wild varieties of game fowls were—

1. Black breasted reds, with fawn partridge hens.
2. Brown breasted reds, with dark brown hens.
3. Red breasted ginger reds, with yellow legs and light partridge hens.

This gentleman made the wild varieties of game fowls his special study, so his opinion should be valuable.

Whatever may be the doubts as to the origin of the game cock, it is certain that he has been bred for his work for many centuries; he is perhaps somewhat changed in appearance, and has been divided into many different breeds and strains by selecting birds that were distinguished by some peculiar mark or colour, and struck the fancy of their breeders; but no one conversant with the subject can doubt the fact that the *original* ancestors must have been possessed of the *true game courage* and *elegance* of *shape*, knowing that it is impossible that any amount of selection should produce truly *game* fowls from ordinary blood.

The first mention of cock-fighting is said to be in the reign of Crœsus, King of Lydia (A.M. 3426), and in India it is mentioned in the "Codes of Manu," written one thousand years before Christ.

At a very early period the cock was famed for his courage; thus Peisthetœrus (Aristophanes "Birds," 483 *et seq.*) relates why the cock was called the Persian bird, and how he ruled over that country before Darius and Megabazus (B.C. 521); the classic poets and historians are unanimous in speaking of his valour, and coins and medals are in existence bearing the representation of two cocks fighting stamped upon them.

The Greeks and Athenians fought cocks. Themistocles, the Athenian General, encouraged his soldiers' valour by pointing out to them two cocks fighting, saying, "These birds fight not for their gods, nor for their children, nor for glory, nor for freedom, but for the sake of victory, that one may not yield to the other;" the example of the cocks having such an effect on the flagging spirits of the soldiers that he led them once more to victory. Afterwards Themistocles instituted a public festival, held annually in the theatre, where the young men were compelled to attend to learn a lesson in courage from witnessing the fighting of cocks and quails. The Rhodian fowls, and those of Media, Chalcis, and Persia, were celebrated for their courage and superiority in fighting, and also for the excellence and delicacy of their flesh.

The Romans, as might be expected, were devoted to the sport, probably learning it from the Greeks, as was the case with many other things. At any rate, they carried it to a great pitch, and fought both cocks and quails. The cock was called "the bird" by the ancients; highly esteemed in some countries, and held sacred in

COCK TRIMMED AND HEELED FOR THE PIT.

others; while, on account of his courage and watchfulness, he was dedicated to Apollo, to Mercury, to Æsculapius, and to Mars. When the Romans under Julius Cæsar invaded Britain they found the fowl in a state of domestication, but forbidden as food and kept for diversion and sport.

One of the first notices of cock-fighting in England is by William Fitz-Stephen in the time of Henry the Second, who alludes to it as the sport of schoolboys on Shrove Tuesday. The school was the pit and the master the director of the sport; indeed attention has been called to the fact that at one or two old-established schools the scholars were being charged with cock-pence quite recently (in one case as much as thirty shillings annually), although the sport had long ceased to exist in the school.

From a very early time the sport has been followed in England without intermission up to the present time, although one or two Acts were passed forbidding it, notably one by Oliver Cromwell, dated March 31, 1654. Most of our kings, however, have been patrons of the sod. King Henry the Eighth, although he passed an Act at one time against cockfighting, built a cock-pit at Whitehall for his private diversion, and is supposed by some to have founded the Cock-pit Royal at Westminster, in which mains were fought regularly until the Act was passed prohibiting the sport in the present century. This building was quite as much an institution as the theatre at the present time—indeed some of the theatres were used for that purpose, notably Drury Lane.

All the Stuarts were fond of cocking. In an old book dealing with orders on the Exchequer in the reign of James the First we find frequently repeated the following order:—"£16 13s. 4d. to William Gatacre, for breeding, feeding, and dieting of cocks of the game for his Highness' recreation."

King Charles the Second is said to have introduced the Pile breed of game, a cock of that colour belonging to his Majesty winning great distinction in the pit. King Charles was passionately fond of cock-fighting, and, according to Sir John Rereby's Memoirs, we find his Majesty's diversions when he visited Newmarket in October, 1684, were as follows:—"In the morning he walked until ten o'clock, then to the cock-pit until dinner time at one p.m.; at three p.m. to the horse-racing on the heath; at 6 p.m. to the cock-pit again for an hour, and then to the play." Not only the royal household, but the whole Court, moved to Newmarket for a month's stay, horse-racing being only one of the amusements. There were also hunting, coursing, cock-fighting, theatrical plays, and hawking, which latter, according to the "Harleian Miscellany," was one of the "Merry Monarch's" favourite diversions, "because it was so commodious for the ladies."

I might go on quoting from old works and writings *ad infinitum*, but however interesting they may be, it is not possible to do so in a little work like the present. It would only be multiplying proofs of the general breeding of the game fowl throughout the length and breadth of the land, including Scotland, Ireland, and Wales. Almost every noble family had their particular strain of cocks, until at the beginning of the present century perfection seems to have been reached. To show in what quantities they were bred, it is said that one breeder alone had three thousand cocks at walk, each cock's pedigree and parents on both sides being known, and each bird marked in foot, nostril, eyelid, or wing, as the case might be. Many people imagine that these birds were of all colours, but it was not so; the different colours were bred much more perfectly than at present, and eye, beak, and leg must match, so that, although there might perhaps be a hundred cocks or more of one strain in the pens for training, yet they would be almost all alike to a feather.

The celebrated winning cocks of the year were as regularly engraved as the winners of the Derby, and as much care was lavished in breeding them; all sorts and conditions of men were represented, and the names of Lords Derby, Sefton, Anson, Lowther; Dr. Bellyse, Halford, Germain and Clarke, may serve to show how general was the pursuit; whilst in the Church itself were many enthusiastic admirers of the sport; one noted Dean of York frequently attended the cock-pit, and bred such fine cocks as to hold his own against all comers. The Rev. Mr. Brooks, of Shifnal, Shropshire, was also a large breeder of good birds. At York the cock-pit was near the Cathedral, and at Canterbury the cock-pit was an apartment of the gateway forming part of St. Augustine's Monastery. The last pit built in England was the one at Melton Mowbray, at a cost of 700 guineas. The old pit at Chester is about 45ft. by 45ft., nearly 40 feet high, with nine large windows, and a glass dome over the pit which was 20 feet in diameter.

Enough, I hope, has been written to show the modern fancier how generally, and in what perfection the game cock was bred by our ancestors, and how long he has been famous in England. Long may he remain the type of beauty and dauntless courage, as he will remain if the modern fanciers and those who can never be satisfied unless by some new departure, do not spoil, transform and *improve* him away.

CHAPTER II.

THE REVIVAL OF OLD ENGLISH GAME IN THE SHOW PEN.

BELIEVE that the first poultry show was held in 1846, and from that time exhibitions of fowls have increased in popularity until they have assumed their present position, and have become an institution in most countries. Probably they have done much good in encouraging the importation and breeding of new varieties; to the game fowl they certainly have done much harm.

When fanciers began breeding game fowls for show purposes they began to destroy their most valuable points, until at the present time the modern show game presents the curious anomaly of a *game cock* that is *not game*, and lacks all the attributes of a *game fowl*. It is not intended by this to decry the modern show production, but simply to state facts. Many, doubtless, admire the modern bird, and some of the old fanciers who have lived to see the game cock gradually resolved into the bird now seen sitting in the show pen, —which on being "stirred up" rises, first on his hocks, then on his feet, then puts his head through the top of the pen, and stands up in all his beauty (?)—must indeed be gratified to know that they have *improved* the game cock to this, a caricature of the heron. Verily they have their reward! Now let those fanciers who take up true game for exhibition purposes take *warning* by *this*. Do not, I pray you, try to improve the true game cock by adding, exaggerating, or making points which true specimens of the breed do not *already* possess. It is, perhaps, less the fault of the breeders than the judges, since it is the judge who awards the prizes, and it is to please him that the exhibitor must breed if he wishes to be successful. A judge should show no partiality for any particular breed or colour, so long as the colour is good of its kind, with eye, beak, and leg to correspond. Nor should mere size be taken much into consideration; strength, activity, good handling properties, keenness, gameness of aspect, cleanness of feet and legs, smartness and purity of feather should be the chief points. The true game cock will not vary from his proper shape if "fanciers" will rest satisfied with the "king of birds," as he is, and which the old breeders bred to such perfection, that modern breeders can never hope to *excel*, if they ever equal them. The game cock as shown at poultry shows being so altered, as already mentioned, in all his chief and essential points, the game classes began to fall off, and ordinary visitors failed to be attracted by a bird which, to all but the eye accustomed to them, must be considered ungainly and practically useless; and about 1882 I fancy the *general public* believed the true game fowl to be a thing of the past, and that he had ceased to exist in this country. Here let me say that it was *solely* among cock-fighters that he did still exist. Many of these had had the same breed for generations, and would neither sell birds nor eggs. They were not led away by showing, or fashion, or money-making. Their requirements were *purity of blood, courage*, heel, activity, strength and soundness of constitution. About the year 1883 a class for the true or old English game fowls was given at Clayton Moor in Cumberland; then Aspatria followed suit, and Wigton provided classes. In 1887 the Old English Game Club was founded, and although this club was given up about two years afterwards, it had nevertheless succeeded in assisting very considerably to bring Old English game prominently before the public—notably, by the classes at Birmingham—and the Dairy shows, which have been since continued and are well filled, forming an additional attraction. The Royal Agricultural Society also gave classes about this time with such success that they are not only continued, but the prizes are now confined exclusively to Old English; and at the present time all the principal shows provide classes for them. The old-fashioned birds appear to be coming quite to the front again in the shows, and our earnest hope now is that they will not be transformed, as their *now* distant cousins, the *modern* game, have been. I think this a very necessary warning. Do let us rest satisfied with the game cock as he is and was two centuries ago. Some of the judges *already* show too much preference for the white-legged breeds, for *wheaten* hens, for more size, and encourage loose feather. For a common table fowl, and a table fowl only, the white leg is good, but it must be remembered that the game cock is something more than an ordinary barn-door fowl. And, as regards size, *I* have *never* seen a *very* large cock so perfectly symmetrical, active and smart as a medium-sized bird; nor is anything to be gained by such large birds either as game cocks or in the production of eggs and chickens. Poultry keepers, from even an ordinary point of view, should bear in mind that great frames and bone are unprofitable, that the fine quality and great proportion of meat, smallness of bone, and *absence* of offal are never combined with great frames, and great frames and great appetites go together, and that very large animals or birds are never so prolific as those of ordinary size, and are also more liable to unsoundness. Wheaten hens, too, should certainly take a second place to the partridge-coloured as a mate for black-breasted reds or for showing.

CHAPTER III.

THE POINTS OF GAME FOWLS.

HE points laid down in the standard drawn up by the "Old English Game Fowl Club," in 1887, are as follows:—

The Old English Game Fowl Club.
Standard of Perfection.

COCK.

Head, narrow, of medium length; beak slightly curved and strong at the base; eyes prominent and bright, with quick and confident expression, and alike in colour.

Muff (if any) to be short and thickly feathered.

Layback (if any) to lie close to head, and extend straight back.

Neck, rather long, and very strong at junction with body; the hackle composed of long feathers, covering the shoulders.

Back, medium length, broad across the shoulders, tapering towards the tail, the saddle feathers long and flowing.

Breast and body, breast broad, full and straight; body medium length, firm and muscular, convex on the sides, broad at the shoulders, and tapering towards the tail.

Wings, long, strong, and inclining to meet under the tail.

Tail, nearly upright, full and expanded; sickle feathers abundant, broad and well curved; main tail feathers abundant, broad, with hard and strong quills.

Legs and feet, thighs short and stout, shanks rather long, of medium bone, flat and clean; scales smooth and close; toes long and spreading, the back toe standing well backward, and flat on the ground.

The whole body to appear symmetrical and muscular, and about an even balance when handled by the sides with the fingers round the thighs.

HEN.

The hen is to be similar to the cock in all points of symmetry, &c.

POINTS OF COLOUR IN BLACK-BREASTED RED GAME.

COCK.

Face, bright red.

Eyes, clear red.

Neck and saddle, dark golden red, free from dark streaks.

Breast and thighs, black.

Back, deep red.

Wings, deep red with a rich dark blue bar across; secondaries bay colour; primaries and wing ends black.

Tail, black, with lustrous green gloss.

Legs, white, yellow, willow, &c.

HEN.

Face, bright red.

Eyes, clear red.

Neck, golden red streaked with black.

Breast and thighs, salmon colour.

Back and wings, partridge colour.

Tail, black, with a shading of brown.

Legs, to match cock.

POINTS OF COLOUR IN BRIGHT RED GAME.

COCK.

Face, bright red.

Eyes, clear red.

Neck and saddle, light golden red, free from dark streaks.

Breast and thighs, black, with an edging of brown on breast feathers.

Back, bright red.

Wings, wing bow bright red; in other respects similar to black red.

Tail, black, with lustrous green gloss.

Legs, white or yellow.

HEN.

Face, bright red.

Eyes, clear red.
Neck, deep golden red.
Breast and thighs, salmon colour.
Back and wings, clay or wheaten.
Tail, black, with a shading of brown.
Legs, white or yellow.

POINTS OF COLOUR IN BROWN-RED GAME.
COCK.

Face, dark red.
Eyes, dark brown or black.
Neck and saddle, rich brown, streaked with black.
Breast and thighs, brown, laced or mottled with light shades.
Back, dark rich brown.
Wings, shoulders dark brown, rest of wings black.
Tail, black.
Legs, dark willow or black.

HEN.

Face, dark red.
Eyes, dark brown or black.
Neck, golden, streaked with black.
Body, black, or of a uniform brown mottle.
Tail, black.
Legs, dark willow or black.

POINTS OF COLOUR IN BLUE-RED GAME.
COCK.

Face, bright red.
Eyes, red.
Neck and saddle, light red.
Breast and thighs, slaty blue.
Back, shade darker red than hackle.
Wings, shoulders and wing bows same colour as back, with a slaty blue bar across, secondaries bay on outer web, primaries black.
Tail, bluish black.
Legs, white, yellow, or blue.

HEN.

Face, bright red.
Eyes, red.
Neck, light golden, slightly streaked with black.

Breast and thighs, light salmon colour.
Back and wings, ginger blue.
Tail, slaty blue.
Legs, white, yellow or blue.

POINTS OF COLOUR IN PILE GAME.

COCK.

Face, brilliant red.
Eyes, bright red.
Neck and saddle, orange red or chestnut red.
Breast and thighs, white.
Back, deep red.
Wings, shoulders red; secondaries bay on outer web and white on inner web, appearing bay when closed.
Tail, white.
Legs, white, yellow, or willow.

HEN.

Face, brilliant red.
Eyes, bright red.
Neck, light chestnut.
Breast and thighs, chestnut shading, lighter towards thighs.
Back and wings, white.
Tail, white.
Legs, white, yellow, or willow.

POINTS OF COLOUR IN YELLOW DUCKWING GAME.

COCK.

Face, red.
Eyes, red.
Neck and saddle, straw yellow, saddle a shade darker than hackle.
Breast and thighs, black.
Back, deep gold colour.
Wings, copper maroon with a glossy green black bar across; primaries black, secondaries white, sometimes tipped with black.
Tail, black.
Legs, yellow, willow, olive, or blue.

HEN.

Face, red.
Eyes red.

Neck, silver, striped with black.

Breast and thighs, salmon red, shading off to ashy grey on thighs.

Back and wings, silver grey free from red or brown tinge.

Tail, dark grey, almost black.

Legs, yellow, willow, olive or blue.

POINTS OF COLOUR IN BLACK-BREASTED SILVER DUCKWING GAME.

COCK.

Face, red.

Eyes, red.

Neck and saddle, silver white, free from dark streaks.

Breast and thighs, black.

Back, silver white.

Wings, shoulders and wing bows silver white, wing bar steel blue, secondaries white on outer web, primaries black.

Tail, black.

Legs, yellow, white, olive, or blue.

HEN.

Face, red.

Eyes, red.

Neck, silver, striped with black.

Breast and thighs, pale fawn.

Back and wings, dark grey.

Tail, grey and black.

Legs, yellow, white, olive or blue.

POINTS OF COLOUR IN WHITE GAME.

COCK AND HEN.

Face, scarlet-red.

Eyes, bright red.

Plumage, all over pure white.

Legs, white or yellow.

POINTS OF COLOUR IN BLACK GAME.

COCK AND HEN.

Face, red.

Eyes, red.

Plumage, all over glossy black.

Legs, black, or dark.

BRASS WINGS.

Same as above, with the exception of a little dark lemon on shoulders of cock.

POINTS OF COLOUR IN SPANGLED GAME.

COCK AND HEN.

Face, bright red.

Eyes, red.

Plumage, either black, red, blue, or buff, spangled with white, spangling as even as possible.

Tail, black and white.

Legs, willow, yellow, or black.

SCALE OF POINTS.*

Head	...	...	...	...	8	Wings 10
Beak	...	...	...	...	4	Tail 5
Eyes	...	...	...	...	4	Legs and Feet 12
Neck	...	...	...	...	10	Closeness of Feather 6
Back	...	...	...	...	10	Colour 10
Breast and Body	...	...	...	15	Handling 6

These rules were, however, drawn up almost entirely by the North Country breeders, and as they are somewhat brief, a further description is necessary.

Beak, big, crooked, hawk-like, pointed, rather short; a long beak, especially underneath, lacks holding power.

Eye, large, fearless, bold, quick and fiery.

Head, small and tapered.

Throat and face, skin of fine quality, loose and flexible, small throat and tight skin causing difficulty in breathing when in violent exercise.

Neck, large boned, round, strong, and of good length.

---

* I have given the above "Standard" on account of its being drawn up at the time of the advent of Old English Game in the show-pen, and showing what North country judges preferred; it is not intended as a guide to the reader, who has the other reliable list of colours to go by. Some of the most glaring errors I have taken the liberty of altering from the original, as they were quite misleading, such as in Brown-reds, having orange or dark red colours, which would turn them into brown breasted orange reds, or brown breasted dark reds, brown breasted reds being of many shades of colour, but true Brown reds have only two colours; the wing tips of Duckwings are also described as black, and white legs approved of in some of the colours where they would be quite out of place.

Back, short, broad across the shoulders, tapering to the tail.

Breast, broad, straight, full, prominent; the pectoral muscles being well developed, gives power to the wings.

Wings, large and long; quills powerful and strong.

Tail, large and spread.

Belly, small and tight.

Thighs, very short, round and muscular.

Legs, strong, clean-boned, not round or gummy like other fowls, nor too upright, nor too wide apart, but parallel with the body, and well bent at the hocks, which is important; spur small and set very low down.

Feet, flat, thin, spreading, long taper toes and nails, the hind toe extending straight back and flat on the ground, not twisted sideways or "duckfooted."

Plumage, the feathers strong, hard, close, sound, glossy and sufficient.

Carriage, bold, smart; the movements quick and graceful; proud and sprightly as if ready for any emergency.

Handling, clever, flesh firm, but corky and light, mellow and warm, with strong contraction of the wings and legs.

A writer on cocking nearly three centuries since says:—"In shape that cock is ever held the strongest, which is largest in the girth, which you shall prove by grasping him about, from the points of your great fingers to the joints of your thumbs, he should be sprightly and prowd in motion; his head small like unto the Sparhawk, his eye large and quick, his beak strong and crooked and big at setting on, in colour suitable to his plume; the beame of his legge very strong and according to his plume." He should carry no offal, lumber, or superfluous flesh, and should be in perfect health and of sound constitution; a poor or weak constitution is totally useless, and the only cure is death. This applies equally to both cock and hen.

The hen should resemble the cock in all points, making allowance for the difference in sex. She should be wide in the back, short-legged, have a small comb, wattles, &c.; be strong, yet clean and bloodlike in feet and legs; strong beak; short wide body; her wings well clipped, almost meeting beneath the tail; her movements should be quick and alert, and she should have a neat and gamey appearance, and if spurred so much the better. It is needless to add she should match the cock in eye, beak, leg and colour.

CHAPTER IV.

THE DIFFERENT BREEDS AND COLOURS.

THE list of breeds and colours here given may be considered absolutely reliable. It was published early in the present century, owing to a disagreement as to terms, &c., when the match bills were made out on the occasion of the main of T. Bourne against Flemming at the Cock-pit Royal, Flemming disregarding the provincial terms of the Cheltenham men. On the weighing night, at a supper given at the " Cock and Tabor," at which many experts were present (among others Nash (2), Varley, Woodcock, Clarke, Martin, Bourne, and Flemming), the matter was discussed, and, to simplify matters, a uniform list of colours was drawn up, discarding provincial names and undesirable crosses, and fixing the originals. It is said that every proposition was assented to or negatived by a show of hands, the only point in which the experts were entirely unanimous was in placing the Black red before all others in merit of purity of blood. The following is the list copied from the original MS. and sent me by a well-known authority on game fowls, to whose kindness I am indebted for much information here given.

LIST OF COLOURS.

REDS.

1. Black breasted blackred—black eyes, beaks, and legs, gipsy faced.
2. Black breasted dark red—dark eyes, beaks, and legs.
3. Black breasted light red—red, yellow, or daw eyes, white, yellow, or carp legs, secondaries when closed chestnut.
4. Streaky breasted ginger red—red, yellow, or daw eyes, white, yellow, or carp legs, wing secondaries chestnut.
5. Brown breasted brown red—dark eyes, dark or bronzy legs.

GREYS.

6. Black breasted dark grey—black eyes, beaks, and legs, wing secondaries black.

7. Mealy breasted mealy grey—black eyes, beaks, and legs.

8. Black breasted birchen duckwings—red or yellow eyes, yellow legs.

9. Black breasted silver duckwings—pearl or light eyes, white legs.

10. Brown breasted yellow birchen—yellow eyes, yellow legs.

PILES.

11. Smock breasted bloodwing pile—red eyes, white legs.

12. Streaky breasted ginger pile—red eyes, yellow legs.

13. Streaky breasted custard pile—red eyes, yellow or white legs.

14. Marble breasted spangled pile—red eyes, white legs.

15. Ginger breasted yellow pile—red or yellow eyes, yellow legs.

DUNS.

16. Dun breasted dun—dark eyes, dark legs.

17. Dun breasted blue dun—dark eyes, dark legs.

18. Streaky breasted red dun—dark eyes, dark legs.

19. Dun breasted yellow dun—dark eyes.

BLACKS.

20. Black breasted blacks—black eyes, legs, and beak, gipsy face.

21. Black breasted brassy winged—black or red eyes, bronzy legs.

22. Black breasted furnaces—dark eyes, dark legs.

23. Black breasted polecat—dark eyes, dark legs.

WHITES.

24. Smock breasted smock, or pure white—pearl eyes, white legs and beaks.

25. Smock breasted white—yellow or red eyes, yellow legs.

26. Cuckoo breasted cuckoo—red eyes, legs various.

27. Spangle breasted spangle red—red eyes, legs various.

VARIETY.

28. Henny—legs, eyes, and beak to match plumage.

29. Muff—legs, eyes, and beak to match plumage.

30. Tassell—legs, eyes, and beak to match plumage.

31. Indian—daw eyes, yellow or willow legs.

I will now make some further remarks on the various breeds generally found at present, and will begin with the black breasted light reds with white legs, as they seem the most popular at the present time with fanciers and at the poultry shows.

BLACK-BREASTED REDS WITH WHITE LEGS.

These birds, sometimes called Lord Derby's, form one of the best strains of game fowls. Lord Derby's breed, however, had daw (grey) eyes, and partridge was the most usual colour for the hens; and although *some* of the red-eyed birds have some of Lord Derby's blood they cannot be considered pure Derbys. Red eyes are necessary under judges at shows at present, but that does not alter the fact that the grey eye was a peculiarity of the Derby strain. Another consists in their having some white or grizzled feathers in wings, or tail, or both; in ordinary black-breasted reds *any* white is a fault, but this breed *always* shows some. No breed of fowls were better in the pit; many a one winning two or three battles, and one was noted as the victor of no less than nine encounters. No breed of fowls excels them for the table; the flesh is beautifully white, short in fibre, and extremely delicate, while there is a great quantity of it in proportion to the small amount of bone; the breast is full and large, and the thighs short and meaty. A celebrated Liverpool physician used to say there was more nutriment in a Knowsley fowl than the largest capon in the London market. Knowsley was the seat of the late Earl of Derby, where this strain was kept for upwards of a century. Thomas Roscoe had sole charge of breeding and walking out the Knowsley birds. They are generally good layers, but as they often want to sit they should be checked if not wanted for that purpose, when they will soon commence laying again.

The weights run about $5\frac{1}{2}$ lbs. for cocks and 5 lbs. for hens. I have seen them much smaller and also very much larger. These are the average weights.

As sitters and mothers, like all game fowls, they are excellent; while the chickens are hardy and soon feather, and are wonderfully quick and active in scratching and searching for insects, &c. The colours should be as bright and the markings as distinct as possible in the cocks, and the hackle bright red above and white beneath, showing a white hackle when cut out in the pit.

When Mr. Harrison Weir visited Knowsley, shortly after the

Earl's death, he found partridge hens and also some wheatens, which were the produce of the grey cross, introduced about 1827. Breeders now use the wheaten hens very much. They must be considered as useful for breeding bright coloured cocks, and, in my opinion, their proper place is in the breeding pen. But so long as judges persist in giving them prizes, to the exclusion of partridge coloured ones, they will of course be shown.

In an article written on the breed by Mr. Weir, last year, he says he knows of no breed he would keep in preference, since they fulfil so many excellent conditions—fine in form, graceful in carriage, beautiful in colour, small in bone, plenty of meat of fine quality, white skin and fat, absence of offal, good layers of excellent flavoured eggs of pretty colour, and seeking their own living far and wide. A wheaten hen, mated with a bright-coloured cock with a *black* breast, will produce bright coloured cocks and wheaten hens; but if partridge coloured hens are wanted they must of course be used to breed from, and the chickens as a rule come more uniform; although the cocks are apt to come a trifle dark but sounder in colour, and the wheatens, if used too much, produce mealy breasted cocks.

Judges object to white feathers in the tail, and although there is no objection on the score of *purity* of *blood*, a cock with much white in his tail is somewhat handicapped in the show pen, although a few of our best judges do not lay too much stress on this point, as they know that the best birds will often show it, and I think it ought not to count more than a point or two against a bird good in other respects. The legs should of course be white with a nice pink shade down the sides, the nails white, and any black on the feet objectionable, the beak white, or striped with horn colour, and for the show pen a red eye is preferred, which should be bright and fiery.

BLACK-BREASTED REDS WITH YELLOW LEGS.

This was a very usual colour in game fowls from eighty to a hundred years since, and several very celebrated strains had legs of this colour. When good they are very handsome in appearance. At the present time a red eye is necessary, a pale one being objectionable and a fault; the cock's breast should be black up to the throat, and the colours bright. They are usually of strong constitution, and run to larger weights than the white-legged birds; most of the hens are good layers of large eggs. This colour is a favourite in

America, where the yellow leg is always admired; the cock should be a good clear bay colour on the triangular space on the closed wing (that is, the secondary quill feathers that show when the wing is closed), and not black or crow-winged as in dark reds; the hens should be partridge colour, and for breeding cocks wheatens are sometimes used, as in other bright reds. These birds are sometimes bred with streaky or shady breast (*i.e.*, an edging of brown on the breast feathers, and the wing bar sometimes is brown), they are then known as shady-breasted bright reds. I have known these colours breed heavier weights than most breeds, sometimes reaching nearly nine pounds in cocks, and seven and-a-half for hens, but these great weights are seldom advantageous; such large birds being generally less prolific and beautifully proportioned than those of medium size. The noted birds of Lord Lowther and Holford were of this colour.

BLACK-BREASTED REDS WITH DARK, WILLOW, OR CARP LEGS.

The black breasted reds with carp coloured legs were a very excellent strain, no breed will come more true to colour; the hens are a good sound partridge hue, free from red on wing; they closely resemble the above variety except in colour of leg and beak, although they were considered at one time not quite so hard in the pit. This seems the most general colour for game in the Midland and Southern Counties, just as the white and yellow legged birds seem most numerous in the North. The colours are usually very pure and bright, and they carry somewhat less offal than some of the Northern birds. The hens are some of the best of layers. Old Sir Harry Goodlake had a fine strain of this colour with dark olive legs. The bright green legs so commonly seen now are probably the result of crossing the dark legged strain with those having yellow legs.

THE TRUE BLACK REDS.

Of black-breasted reds, as mentioned above, there are many variations, from dark red to the brightest orange hues, but of the true blackred, or black-breasted blackred, rather, there are two colours only, viz., black and red. In colour then the black-breasted blackred should have the hackle, back, scapulars, and shoulder coverts of a clear, vivid dark red, free from any black on the surface, but black at the *roots* of the feathers, while the breast, thighs, belly, tail, primary and secondary wing feathers, should be pure black. The brood hen for such a cock should be a dark partridge colour, bright red hackle above, black beneath, clean brick breasted, and such to the posterior. In both sexes the eye, beak, and legs are black, and the face that dark red, which is known as "gipsy face." These are supposed to be the purest strain of game fowls, and no birds had a higher reputation in the pit. It was from this breed that Dr. Bellyse produced his celebrated brown-breasted reds, by crossing them with the piles of Cheshire.

BLACK-BREASTED DARK REDS.

This strain is closely allied to the above, the principal difference being that the eyes, beak, and legs were dark, instead of black, and the hackle generally striped with black; the flight feathers are all black or "crow-winged."

Sant, a Derbyshire breeder, noted for breeding, and supplying a large number of good cocks, bred mostly birds of this colour; they are described in a work, published at the beginning of the century, as "a very dark blackred, striped uncommonly black upon the neck, black beak and legs, very lofty, and fought at high weights. The dark reds of Col. Mellish were also noted about that time.

STREAKY BREASTED GINGER REDS.

These birds differ from the black-breasted in not having the blue bar across the wing, which instead is of a light bay, the breast is streaked with brown, and the hackle and saddle a ginger red; the hens are a light reddish colour, eyes red, yellow, or daw, and yellow legs.

BROWN BREASTED BROWNREDS.

The above is one of the greatest favourites with sportsmen, indeed, I may say that after black breasted reds, they are generally the most popular. Dr. Bellyse, who throughout his long life was invincible in the cock-pit, produced them by crossing the true black red with the yellow legged piles of Cheshire. The cock should be a rich brown in breast, hackle, shoulders, and saddle; tail and primaries black, the hackle striped black. Sometimes the hackle is a rich orange colour, striped with black, and dark orange shoulders, they are then known as brown breasted orange reds; the brown breasted dark reds have dark red hackles, &c. There are many varieties of breast colour, some prefering the copper breasted, others streaky, and others, again, the laced breasted birds, but these all differ from the true brown breasted brown red. The hen dark brown, or blackish throughout, except the hackle, which is dark golden, striped with black; gipsy or dark red faces, and dark legs, eyes and beaks are necessary in this breed. The late Mr. W. Cobden's birds were of this colour, and I know of one, a hen, now preserved in a glass case, that moulted a spangle, throwing back to the original piles, and then became white and the legs turned yellow after she had moulted two or three times, which was of course a sport liable to occur in all breeds of animals and birds.

THE BLACK BREASTED DARK GREYS

Closely resemble the true blackred, saving that the red gives place to a dark steely grey; sometimes they are striped in the hackle, the hens to match them are very dark, with grey striped hackles.

THE MEALY-BREASTED MEALY GREY.

The author of the "Cocker" describes these birds as follows, "A clear mealy grey, nearly white breasted, without spot or streak, 4 lbs. 6 oz. to 4 lbs. 8 oz., high standing, bony and black legs, close feathered, short hackled, small snake head and full dark eye, their walk easy, firm, and majestic, their breasts gracefully prominent, shoulders broad and up, their bodies gradually tapering to the tail; their actions were in unison with their shapes."

BLACK-BREASTED BIRCHEN DUCKWING.

Many of our readers are no doubt acquainted with the two prints by Marshall of a cock in full feather and one trimmed for the pit of this breed. It is supposed they were bred from the black-breasted red, the yellow birchen and the grey duckwing hens. They are beautiful and good birds and have usually very good strong constitutions; yellow legs and red eyes and a dark birchen colour across the shoulders and back are necessary in the cock; in other respects he much resembles the yellow or golden duckwings.

BLACK-BREASTED SILVER DUCKWING.

This handsome strain was known many years ago as Lord Hill's; it was these birds that were crossed with the Dorking, producing the present strain of silver grey in that breed; they were bred with pearl or light eyes, but red eyes are preferred by most breeders now; the cock is a glossy black in breast, thighs, belly, primary wing feathers and tail, with a blue bar across the wing; in hackle, shoulders, saddle and secondary wing feathers, a clear silvery white; the hen a beautiful blueish silvery grey, marked with darker pencilling, quite free from any rusty feathers on the wings, the tail dark and hackle silvery white striped black, legs and beaks are white and the breast is fawn colour; no strain will breed more true to colour than these.

YELLOW BIRCHEN.

The brown-breasted yellow birchen with yellow eyes, beaks, and legs, are not often seen now; they are hard, gamey-looking birds, the hackle and saddle feathers a pale straw colour, having a shade of birchen showing thoughout; the wings brown, as are also the secondaries, wanting the bar across and white space, as seen in the Duckwings. Mr. Nunis' strain of Yellow Birchens, known as the butcher breed, from their killing qualities, held their own for some years in the early part of the present century.

YELLOW OR GOLDEN DUCKWINGS.

This is a very handsome variety. The cock should have a clear hackle, free from ticks or stripes, and of a bright straw colour; saddle feathers one shade darker, shoulders should be a bright yet deep golden colour, the wing has a clear steel blue bar across it, the triangular space on the remainder of the closed wing being pure white; breast, thighs, belly, tail, primary wing feathers and tips of the secondaries black; the legs yellow or willow. The hen resembles the silver Duckwing hen in colour, except that the appearance is not quite so silvery and the breast is a deeper fawn colour; the hackle white, striped lightly with black, tail shaded to black, any rusty hue on the wings is very objectionable. Yellow Duckwings are frequently bred from a bright-coloured black breasted red cock and good coloured Duckwing hens, and a silver Duckwing cock is sometimes used for bringing back the pure colour in pullets. The Isle of Wight Yellows, so famous in days gone by, are all now extinct, although a few strains have a dash of their blood.

PILES.

These birds have been notorious for their winning powers in the pit, and are favourites with most sportsmen; while the handsome colours, together with their good laying qualities, should recommend them to the fancier. It is said that the colour was introduced by King Charles II., whose pile cock was very celebrated, they usually had the credit of carrying a very deadly heel, and without having quite so much gameness and staying power as some strains, were exceedingly quick and dangerous to their opponents. The Earl of Derby had a bird of this colour that won many battles; while the Cheshire Piles, with their peculiar mode of fighting, bore off most of the honours of their day. The Mansell Piles too were very famous also about that time. Pile game, like many other breeds, were perfected by the closest in-breeding; even the dark and light varieties of the Cheshire Piles were kept to themselves, and when crossed with other pile coloured birds generally deteriorated.

SMOCK-BREASTED BLOOD-WING PILES.

These cocks should resemble a bright coloured black breasted red, except that the black is exchanged for white, and the bar across the wing white. The hen is golden yellow in head and neck, the latter striped with white, salmon red breast, the colour of back and wings a clear creamy white; frequently the shoulders are marked slightly with red. This for exhibition is perhaps the handsomest colour in piles, and they can with care be bred very true, occasionally a black breasted red cross is necessary for these high coloured ones, but not nearly so often as supposed. Black breasted red cocks and pile hens usually throw bright coloured cockerels, while good coloured pullets are sometimes bred from a pile cock, and hens bred from a pile cock and blackred hen. White cocks and black breasted red hens also throw piles.

STREAKY BREASTED PILES.

Some breeders prefer the streaky breasted birds. The streaky breasted ginger pile is as described—that is with breast streaked red, and the hackle and shoulders and saddle feathers a light *ginger* red, legs should be yellow. The streaky breasted custard pile is even lighter than the above, and the white ground should have a creamy appearance; the hens of these varieties should be of the same colour as the pile hen previously described, but correspondingly lighter to match the cocks.

THE MARBLE BREASTED SPANGLED PILE.

The breast in these birds should be marbled, or spotted with red or sometimes black; the hackle and saddle red, white at roots of feathers, with a few tips of white (and sometimes a little black), shoulders red, tail white, ticked more or less with black; these birds have a parti-coloured or piebald appearance, and to admirers of those colours are handsome; the hens are a fawn colour speckled with white, and both sexes have white legs and beaks, and red eyes, the white spangles should be as even as possible throughout.

GINGER BREASTED YELLOW PILE.

The breast ginger or tawny in colour, yellowish red hackle and saddle, red shoulders, white flights, and tail mostly white. These are rarely bred now, and are not so handsome as some of the other varieties; yellow eyes, beaks and legs. These are considered by some the original piles.

DUNS.

There are no more celebrated strains of fowls than the various strains of Dun. Sir Francis Boynton, Mr. Elwes (one of whose red duns won the extraordinary number of twenty-seven battles), and that great cocker and soldier, Col. Mellish, bred some of the finest ever seen. They are particularly hard and workmanlike in appearance, and the hens have a very neat, trim, and gamey look.

The dun breasted dun, with dark eyes, legs, and beak, seems to require no further description. The colour should be level and uniform, and the hen whole coloured to match.

The dun breasted blue dun, the hackle, saddle, shoulders, and primaries should be a dark blue, the remaining portion dun; the breast is sometimes laced with dark bluish edgings to each feather. The hen should be dun, the hackle a shade darker than the rest, and is sometimes laced throughout with dark blue edgings to the feathers. These laced birds, when of good, sound colour, should be better known, as they are extremely pleasing.

The streaky breasted red dun is perhaps the most popular of all the birds of this colour; cocks—beak and legs dark, eyes *dark* red, breast slate colour, with edging of gold on each feather, or striped gold colour; hackle golden red, shading golden towards the bottom, striped with dark stripes, shoulders clear darkish red, saddle feathers red gold colour, striped with slate, tail and secondary wing feathers slate colour. The hen should be slate colour, laced with gold on breast, and having a golden striped hackle; eye, beak, and legs to match cock.

The dun breasted yellow dun is a dun fowl, with light yellow, hackle and saddle, and reddish yellow shoulders; in other respects dun.

BLACKS.

The black cocks of Lord de Vere were noted, and continued in great notoriety when afterwards bred by Mr. Thomas Wilson, of Burton, Staffordshire. He describes them as " a perfect jet black, gipsy faced, black legs, rather elegant than muscular, lofty in their manner of fighting, close in their feather and well shaped." True Blacks require no further description than black entirely free from any other colour, and having black beaks, legs, and eyes, and gipsy faces.

Black breasted brassy winged are the same as above, except that the cocks have a patch of flame colour on each wing. They often have red faces and eyes, and their legs should be bronze.

Black breasted Furnaces, with dark eyes and legs, have golden coloured feathers across the shoulders, vary in shade, and are supposed to resemble a furnace; the colour is supposed to resemble flames of fire rising from a coal or black ground colour; black breast and saddle, body fiery red, often tinging hackle and wing, and some of the hens often show a little gold on some of the shoulder feathers; in other respects black. These birds were sometimes bred with bright yellow legs.

Black breasted Polecats.—These are black, but have more red than Furnaces and of a lighter shade, sometimes extending to the hackle and saddle. The hens are dark, or black with hackle to match the cock.

WHITES.

Some of the whites possess a longer pedigree than most breeds of game. The names of Molineauxes, Cholmondeley's, and Raylances were well known to old breeders for the excellence of their white breed of games. The purest strains had white legs and beaks and pearl eyes, but many of the whites now have red eyes, and sometimes yellow legs, which is quite allowable.

They are usually small, and most white fowls *appear* softer in feather than some others at the present time; care in breeding would probably improve them. They were usually known as "Smocks."

CUCKOOS.

The cuckoo-breasted cuckoos are blueish-grey in colour, banded across the feathers with darker or lighter shades, something similar to the colour seen at poultry shows in Plymouth Rocks or Cuckoo Dorkings. Most of the birds I am acquainted with of this colour are north country bred, they should have red eyes and generally white or yellow legs, the hens of course resemble the cocks in colour; the markings, &c., should be as even as possible throughout the plumage, in different birds the colour varies, some being almost white and others blueish-grey in ground colour, while the markings vary from blueish-grey to black. Yellow cuckoos have a buff ground colour.

Spangle-breasted spangle-red, or pheasant-breasted pheasant-reds, as they were often called, are, I believe, almost extinct or bred very rarely at the present time. They were fowls of very attractive and brilliant plumage consisting of red and golden-red spotted and marked with black, almost after the fashion of a "Golden Spangled Hamburg," of whom they are said by some to be the ancestors, and some writers say that the "Golden Mooney" and "Pheasant Malay" owe their existence to the same source. Their markings were very attractive, but from being less successful in the cock-pit they gradually became neglected, although if any be in existence now, they would probably become popular in the show-pen on account of their plumage.

HENNY GAME.

The remarks given here on "henny game" are from the pen of the greatest breeder of that variety at the present time, a gentleman to whom I am indebted for some of the most interesting information contained in this little work.

"If there is any truth in the theory of our game breeds coming from the wild fowls of India, then the Ceylon jungle fowl, being partly hen-feathered, suggests itself as a likely ancestor, besides some of the pheasant tribe. On the other hand we had game fowls long before having any intercourse with that country. Gilbert and other officers have seen henny-cocks fought in India. They were formerly very plentiful in Wales and Cornwall, and there is an old account of a main, fought at Pontefract in the year 1670, hen-cocks *v.* long-feathers. The black hen-cocks of Wales were thought a fit present for a prince, and Pembrokeshire once challenged all England with them. Hunt brought these black-thornes (hen-cocks) into Somersetshire, and old cockers used to tell how when there were no roads for carriages, they carried their hen-cocks on horseback to Exeter to fight. It will be seen, therefore, that hennies are an old-established and well-known variety of British game fowl, from which they differ chiefly in length, form and brilliancy of feather, the plumage of the male bird invariably resembling that of the hens, hence the name of hen-cocks or hennies, and the more rounded, short and free from sheen or gloss they are in hackle, cloak, and tail, in short the more hen-feathered they appear in neck, wings, body and tail the more they are entitled to claim purity of breed; they are generally light in bone, and on that account were never favourite match cocks with the old feeders at the 'Cockpit Royal,' who preferred cocks with more bone, having light corky bodies that appeared larger than their actual weight in the scales; although this was sometimes more than compensated for in the hen-cock's watchful eye, wary posture and sudden rapid onslaught, in fact, like the famous Cheshire Piles, they seem to have an hereditary style of fighting peculiarly their own. It is an undoubted fact that hen-cocks generally get the first fly, and being invariably good heeled cocks frequently crippled or wounded their antagonist, which being effected they were seldom slow to finish. It is admitted this frequently occurs through the apparent apathy of their opponents, who appear disconcerted by their feminine appearance and are unable to decide whether their mission is one of love or war, until pierced with the silver spur. During a long and extensive experience with these birds, I have seen scores of cocks crippled and not a few killed outright before assuming a fighting attitude, while totally off their guard, and even whilst mantling, their death thrust has come upon them sudden as an electric shock. Nothing equals the first dash of two hen-cocks when pitted; we have seen both *stone dead* within *thirty seconds* of leaving the setters' hands—an almost incredible fact to those who have witnessed the violent struggles and muscular contortions of fowls for minutes after decapitation. Fast and furious as they fight a winning battle or kill a sinking cock, they are frequently found wanting in stamina or physique to finish or win in a long and severe contest, thus proving their further resemblance to hens in constitution as

GROUSE COLOURED HENCOCK, THE PROPERTY OF MR. JOHN HARRIS, LISKEARD.

well as feather. They are of various colours, the most esteemed being dark partridge, red-grouse, wheatens, greys, blacks, duns, whites, and spangles; a well-known poultry judge once stated he had seen and bred black-breasted reds, but certainly these were not hennies. The colour of eye, beak and legs should correspond as near as possible with the plumage; having been kept more free from crosses than other strains of game, they retain the original characteristics of the breed, generally run smaller, but sometimes weigh as much as seven or eight pounds, and probably carry more meat with less offal than any breed whatever, which for quality, richness of flavour, delicacy and nutritive properties is second to none. They are the most prolific layers of the whole game race; they cannot lay any claim to that beauty of plumage so conspicuous in other strains of game, and this seems their only drawback, for what they lack in ornament they make up in utility, and for the use of the cottager or those who breed fowls for their own table they are worth all the *new so-called best* breeds that have been introduced of late. There has been a great deal written in the public press on the enormous quantity of eggs and poultry imported, and we need scarcely wonder at this, when the whole country is filled with huge useless fowls, each of which will eat as much food as three of our old English breeds and not lay any more eggs; the cottager cannot afford to buy food for such 'gormandising machines,' so throws up poultry-keeping in disgust. If we were to return to rearing our old English game for quality, Dorkings (not the *great show* Dorkings, but the pure old breed) for quantity, and Minorcas or Hamburghs for eggs, which will pick up enough to live on where Brahmas or Plymouth Rocks would starve, we should soon get more eggs and better quality of market poultry; the Asiatic breeds all have too much waste and offal, and wherever there is immense size there is extreme coarseness to correspond, there being as much difference in old English game and Asiatic as in the best and coarsest cuts of beef in the London markets, yet the public pay the same price for poultry whether good or bad, although there is scarcely any meat that differs more in quality."

MUFFS.

Of muffs there are almost all colours; their only difference consisting in a thick muff or growth of feathers under the throat in both cock and hen. They generally run large, and are strong in constitution, while they have a singular appearance. They are chiefly bred in Cumberland and in the North, and are usually excellent layers and good game birds.

TASSELLS.

Some of the very best birds we have in England are known as "tassells;" that is, they have a small tassell or tuft of feathers on the top of the head behind the comb. In the cock it varies from a few long feathers, extending straight back, to a small tuft. The hens have a more pronounced topknot, sometimes as large as a walnut, of round shape, and the feathers standing nearly upright, but bending over towards the back and giving them an animated and spirited appearance. An old writer on game fowls, from whose book most of the more modern ones were taken, and who wrote some two centuries and a-half ago, says, "The best hennes are the tufte hennes." Ralph Benson's famous reds were of this breed. By a few they were called red-duns, but they were really brown reds, with less black than the more modern stamp. They were strong-boned, good birds, and noted for their game qualities; indeed, all the tassells now kept are noted birds in the pit, getting more savage and revengeful on being struck. The best known strains are black and brown breasted reds with dark eyes, beaks and legs. They are generally excellent layers.

CHAPTER V.

THE SPORT OF COCKING.

I TRUST I have written enough on the various breeds of game to afford my readers who, are not already conversant with them, some information as to their chief characteristics. There are many additional and provincial *local* names of the breeds here mentioned, but were I to add them it would only cause confusion, without effecting any useful purpose. It will be noted that the colour of the breast is always given. The reason was that in old times when cocks were trimmed for the pit the hackle, tail and other parts were cut or trimmed, but the breast was left intact. Hence the birds were described as black-breasted blacks or smock-breasted smocks, as the case might be.

A few further remarks on cock-fighting, as formerly carried on, are necessary. After which we will consider the useful qualities and general management, &c., of *game fowls*, which will doubtless be of more interest to the general reader. Yet a monograph on the breed would be incomplete without some description of the sport.

It has already been stated at what an early age cock-fighting was practised in Britain; it has been proved that the Romans followed the sport here, as metal spurs for cocks have been found, with other Roman remains, in Cornwall, and I believe that there is a specimen in the British Museum, and it is thought that the ancient Britons themselves practised the sport in some form previous to the Roman invasion. It has continued as one of our English sports up to the present time. Many cock-pits are still in existence that were built especially for the sport, though they are now turned to other purposes; one being a Dissenting Chapel, another a confectioner's factory, a third a theatre, and so on. "The Sod" has generally gone hand in hand with "The Turf." Newmarket was a great centre of cocking, and of course, London,

Staffordshire, Lancashire, Worcestershire, Cheshire and Devon, were also great counties for the sport; the cock-matches were noted equally with the races, and "Hebor's list of horse-races and cock-matches" was published annually at Newmarket some eighty years ago. The races were put on one side until the main of cocks had been fought, while at the Quarter Sessions there was generally a main fought ere the magistrates returned to their homes, and such were the amounts staked, that at one main fought at Lincoln, the

THE EARL'S PILE.
Winner of many Battles.
(From a print in the possession of Mr. Jno. Harris, Liskeard.)

stakes were £1,000 each battle, and £5,000 on the odds or main. Gwenep pit is the largest in England, where it is said cocks were fought in spurs of silver and also of gold.

Spurs were long ago known as gaffles or goblocks, and were made of iron, brass, or silver. The silver spurs used had a portion of copper in their composition and were much stronger and more elastic than the finest steel. There is no maker of silver spurs in England at the present time. All matches and mains of importance were fought in silver, they being not quite so immediately destructive as steel; there was time for the birds to show their powers of endurance and their gameness more fully. The shapes varied somewhat in their curves, &c., but the twisted heels, slashers (like a two-edged sword), and three-edged spurs, as used abroad, were not allowed under English rules; some objecting to drop sockets,

BLACK-RED GAME COCK.
Gold Cup Winner at the Cock Pit Royal, 1818.
(From a print in the possession of Mr. Jno. Harris, Liskeard.)

that is, the blade starting below the socket, thus throwing the spur lower on the leg; but the law in reality simply said, "in fair reputed silver spurs." Foul spurs were sometimes made, that is of steel, but with a thick coating of silver covering them, except at the point, which was only just thinly plated.

The art of heeling a cock consisted in setting the spur on a line with the outside of the hock, in the same direction as the natural spur; if too far towards the outside the blows would be ineffectual, while if set too far in, it would cause the cock to cut his own throat. The spurs required to be padded firm at the socket, and tied on so tightly as not to move (since if they came off in the fight they were not allowed to be replaced), at the same time if too tight they would cramp the cock.

In breeding birds for the pit several points had to be attended to; courage, of course, was of the greatest importance to fight to the death, but this alone was useless, unless the cock was a good heeler; the piles and some other breeds were noted for having deadly heels. The cocks also required to have good mouths, for although they should not take hold with the beak *early* in the battle, when they are a little weary, it enabled them to give strength to the blow. Another requisite was that they should "come to every point," that is that they should take hold and strike at any part of their opponent within reach, be it wing, tail, or any part; a hasty manner of fighting *sometimes* denotes a want of bottom or gameness. Shifty cocks, although they sometimes won, were not admired; the action in fighting should be rapid without hurrying, quick, but cautious and wary, to break well with his adversary, that is to parry or ward off the blow, and then hit; since when they both hit together a thigh or wing is often broken. A cock should always press his opponent when he has the advantage, never letting him rise when he has him down.

Cocks varied much in weight in proportion to their size, some being heavy-fleshed, and their bones solid and heavy; while others, even larger, weighed much less from having lighter quality of bone, and being more corky and light in flesh, and these were generally preferred for matches, most cockers thinking a big cock to his weight desirable.

The cock must have been allowed to run as master for sometime previous to fighting; and a good walk where he obtained plenty of exercise and could not be annoyed by other fowls was required, with not too many hens. Previous to fighting he was placed under the care of the "Feeder," in order to get him in the best possible training and condition, as is now the case with human athletes, race-horses, or grey-hounds, and quite as much art and experience are required in the one case as in the other; and such names as Fisher, Wading, Bromley, Potter, Gilliver, Nash, and Varley, with Parker,

Woodcock, Martin, and others are remembered as adepts in the art; nearly all these had systems peculiarly their own, and the directions for feeding and treating cocks in training are many and various. Extraordinary things were given them and most curious compounds.

Bread made of fine flour, oatmeal, and the whites of eggs, with a little cinnamon, was perhaps the simplest; while some cock breads contained nearly a score of ingredients. Peameal was also used and pearl barley; while such things as barley-sugar, hempseed, aniseed, carrawayseed, ginger, rhubarb, yeast and anna were recommended as foods or drugs in various receipts. The following were the lines on which most birds were treated, although minor details differed in almost every case. The Feeder carefully examined each cock that was sent to him to see that he was in perfect health; he judged this by his looking red in the face, his crow being clear, and his feathers being glossy and tight, and his flesh feeling firm to the touch, and by the length and sharpness of his toe-nails, if he had been on a good walk. If considered fit his spurs were sawn off to about half an inch in length, his tail somewhat shortened and he was placed in a pen. These pens were about three feet high and some two feet square, made of board, with two spars in front to admit light and air, and to enable the cock to put his head through to get at his food in the trough attached to each; they were placed in rows about three feet from the ground. The cocks were then usually given some physic, such as rhubarb, senna, or jalap, and covered up close, or sweated, they were then sparred, each pair of cocks having muffles, similar to pads or boxing gloves, tied over their short spurs to prevent injury, about two or three sparrings were usually given in the course of their training, this not only got them into practise, and improved their wind, but also rendered them eager, so that they would commence fighting at once on being put down, since in a match a cock that walked round his opponent and crowed, &c., would probably be struck by the other cock before he had begun fighting. Too much sparring had a contrary effect, and would destroy their courage. They were fed sparingly on nourishing food for some eight or ten days, and were lightened of all superfluous flesh and weight by physic, feeding and exercise, and were weighed three days before the main. After weighing they were fed more liberally, and it was lawful to increase their weight again by any kind of feeding. Sometimes large cocks were reduced very much to fight within the articles of agreement, and after weighing were recovered by jelly, eggs, isinglass, and all kinds of strengthening foods. About fourteen to sixteen ounces is as much as any cock should lose, and some will require to gain rather than lose weight in their training. On the weighing day the match bills were compared, in which all the marks of the cocks, their colours, and full description were taken together with their weights.

The following is an agreement for a cock-match:—

Articles of agreement made the   day of   181— between W. S———— and J. C————.

First, the said parties have agreed that each of them shall produce, shew, and weigh, at the   Cockpit   on the   day of   next, beginning at the hour of   o'clock in the said morning   cocks, none to be less than three pounds six ounces, nor more than four pounds eight ounces, and as many of each parties' cocks as come within two ounces of the other parties' cocks, shall fight for   guineas a battle—that is   guineas each cock, in as equal divisions as the battle can be divided into as pits, or days' play, at the cockpit aforesaid; and that the parties' cocks that win the greatest number of battle-matches out of the number aforesaid, shall be entitled to the sum of   guineas as odd battle money, and the sum is to be made stakes into the hands of Mr. ——before any cocks are pitted, in equal shares between the parties aforesaid; and the parties further agree to produce, shew, and weigh, on the said weighing day,   cocks, for bye battles, subject to the same as the main cocks before-mentioned, and those to be added to the number of main cocks un-matched; and as many of them as come within one ounce of each other shall fight for two guineas each battle, to be as equally divided as can be, and added to each pit or days' play with the main of cocks; and it is also agreed that the balance of the battle money shall be paid at the end of each pit or days' play; and to fight in fair reputed silver spurs, and with fair hackles, and to be subject to all the usual rules of cock-fighting as is practised in London and Newmarket; and the profit of the pit or days' play to be equally divided between the said parties, after all charges are paid and satisfied, that usually are thereupon. Witness our hands this   day of   181—.

Witness   W. S.
J. W.   J. C.

RULES FOR MATCHING AND FIGHTING COCKS IN LONDON.

To begin the same by fighting the lighter pair of cocks (which fall in match) first proceeding upwards to the end, that every lighter pair may fight earlier than those that are heavier.

In matching (with relation to the battles) it is a rule always in London that after the cocks of the main be weighed the match bills be compared.

That every pair of equal or dead weights are separated and fight against others provided that it appears that the main can be enlarged by adding thereto either one battle or more thereby. The above were some of the rules and regulations required in the sport of matches. However there were various kinds.

A cock previous to fighting was "cut out of feather," *i.e.*, his hackle trimmed short, the sickle feathers all cut from the tail, leaving only the fan or straight tail feathers, which were shortened about half; the ends of the flight feathers cut off, the ends of the long saddle feathers, and the feathers round the vent and belly cut close; the artificial spurs being then tied on the bird was ready for action. It is almost needless to say that throughout their training cocks were kept scrupulously clean, with fresh straw in their pens, their beaks and feet washed daily and their legs rubbed. The handler required a quick eye and hand and a steady head, to take in and make the most of every chance in the battle.

Of mains the most usual was the short main, *i.e.*, to show, weigh and fight (say) twenty-one cocks on each side; the stakes being so much each battle and so much on the odd battle or main, an odd number being always provided to prevent the contingency of a draw—which sometimes, however, happened (in the last battle both cocks have been struck dead at the same time). At the Cockpit Royal at Westminster sixty-three pairs were usually shown. They were weighed three days previous to fighting; the colour of breast, eyes, beak and legs taken, as well as colour of body, and any peculiar or distinctive marks they might have, to prevent the possibility of their being changed in the match. All that weighed within one ounce of each other were matched and divided into three or six days' play, the lightest pair commencing the fight. This was the long main.

The Battle Royal is proverbial. In this any number of persons stake a certain sum, and each produces one cock under a given weight, when the cocks are all put in the pit together, the last survivor taking the whole stakes.

In the "Welsh Main" sixteen cocks are matched in pairs, the nearest weights being matched together; the eight winners are then fought, and so on. The ultimate winner having to fight four battles. This is the severest trial of all. Immense sums were formerly staked on these mains. Indeed, at one fought at Lincoln, not so very long ago, the stakes were £1,000 each battle and £5,000 on the main (or odd battle).

At two years old a cock was considered in his prime for fighting, previous to which he was taken up from his walk and handed over to the care of the "feeder" (or trainer). These men were considered as important as the trainer of race-horses is to-day, and each had a system more or less exclusively his own, which he believed superior to all others. Their duties were to get the cock as light as they could without injury to his health on the weighing day, in other words, to produce as large a bird to his weight as possible, at the same time to keep him in the very highest degree of health, spirit, strength, and wind. After the weighing it was their aim again to increase his weight. Their art was to simply divest him of fat and superfluous flesh in the same manner in which racehorses or athletes are prepared at present. Physic, sweating, exercise, were all used to attain this end, and as no cock can remain at the very top of his condition for long together, they used every endeavour to arrive at this point on the day of battle. It was in this particular that the best feeders especially showed their superiority; and when it is remembered that each cock would vary somewhat in condition and constitution when he was first put in the feeder's pens, it will be seen that this was no easy matter. There are many rules and receipts laid down for feeding cocks, and many curious compounds were sometimes given them, the receipts of each feeder varying to such a degree that it seems unnecessary to give them here. Judgment was the chief thing; some birds requiring to lose much weight, others rather to be increased than reduced. They were usually fed for nine or ten days before fighting, before which event they were trimmed or "cut out of feather," that is, the hackle was cut off short, the wing feathers cut from the last sloping feather downwards, the feathers of the belly and vent trimmed short and the long sickle feathers cut from the tail, leaving only the vane, or fan, which was shortened about half. The cock's natural spurs were cut off about half an inch from the leg, and the silver or steel spur tied on, the socket fitting over the natural spur, and the point extending back on a line with the outside of the hock (or rather outside the line of the natural spur). Much art was supposed to be necessary in putting on these weapons. It is mere ignorance which would describe them as cruel. With natural spurs the combat lasts far longer; whilst with artificial spurs the longest fight is over in a few minutes. They therefore actually lessen the sufferings of the birds by *shortening* them. Silver spurs were used by the Romans, specimens having been found in Cornwall and elsewhere amongst Roman remains. They were somewhat different in shape from those now used. The curves and shapes of spurs were arrived at by mathematical calculation, and the quality of the metal must be the most exquisite, as nothing but the most perfectly tempered metal will stand the shock of the gamecock in striking, a bird having been known to drive his *natural* spur an inch deep into a solid board. The art of making silver spurs is now almost lost, and is exceedingly laborious and expensive, the most valued being old ones, as being stronger and lighter than those made now. Spurs vary from one inch to three and a half inches in length.

The laws of fighting, put shortly, were as follows. The setters having put down their cocks six feet apart, kept back, and were not allowed to take up their bird unless one cock was fast in the other or in the pit, or hung in himself, in which case they might be handled and brought to the centre of the pit; if the bird was thrown on his back it was lawful to turn him over only, but removing feathers from beak or eyes was not generally allowed. If from blindness or any cause the cocks cease to fight the law is told, that is, twice twenty is counted, when they are handled and set again: this telling the law is repeated as long as both cocks fight, but ten only is counted at each interval after being put together, either ceasing to peck is told out by a person counting twice twenty, they are then breasted beak to beak, and if still refusing ten is counted and once refused announced and so on until he has refused ten times when he loses; this is the long law. Should both be disabled and refuse to fight before the long law begins, it is a drawn battle. Should both refuse fighting during the counting the winner is the cock who fought last, but should he die before the counting is finished he loses the battle, notwithstanding the other did not fight within the law. The short law is told by a person counting audibly twice twenty, and afterwards asking three times, Will anyone take it? If no one accepts the challenge, the cock is beaten. If, however, it is desired to stop this counting out, the cock may be, in the language of the pit, "pounded;" when he must fight till death, and sometimes unexpectedly recovers and wins. Some cocks are so savage as to be useless in the pit, as they will turn on their handler.

Such are the outlines only of the sport of cock-fighting as formerly carried on in this country. With regard to the cruelty attached to the sport, it must be remembered that no man can encourage or force a cock to fight against his inclination, nor stop him running away if so inclined, that artificial spurs lessen their sufferings by shortening them. The persons who generally exclaim most against the sport are those who are not conversant with the nature of the game cock, and have never seen a cock-fight, they must therefore be incapable of forming a correct judgment on the subject. It must also not be forgotten that the game cock was reared in every enjoyment and luxury till two years old, and then had a fair chance of his life; whereas an ordinary chicken is reared until some five or six months old, perhaps crammed, and finally killed, not always by any means in as merciful a manner as the game cock. It is not desired to write a defence of the sport, but simply to put both sides of the question before the reader, who will I trust approach the subject with what one of our greatest statesmen calls "an open mind."

HEADS OF MUFFED GAME.

HEADS OF TASSELL GAME.

CHAPTER VI.

THE USEFUL PROPERTIES OF OLD ENGLISH GAME.

ENOUGH has already been written of the history, points, and fighting of game fowls; let us now turn to what is of more interest to the general fancier, viz., their useful qualities. It has already been stated that old English game stand in front of any variety whatever as table fowls; carrying, as they do, the largest proportion of white, delicate, and nutritious meat, with the smallest amount of waste, offal, or bone. They carry more breast meat in proportion to their size than any other fowl, and killed from a good run they more than other fowls resemble the pheasant in flavour. Many strains of game are excellent layers of good sized and beautifully flavoured eggs; hennies, black breasted reds, and piles, standing first in this respect. If checked when wanting to sit they will continue laying for some time. I have had game hens lay over sixty eggs before becoming broody, and continue laying throughout the greater part of the year.

As sitters and mothers they are invaluable, quiet on their eggs, close sitters, regular in leaving and returning to their nest, and when the brood is hatched they are the best of mothers, not trampling and killing their chickens by treading them under foot, as is often the case with large and heavy breeds; whilst against cats, vermin, or indeed any enemy, they will defend them to the last. Another recommendation is that, according to their size and weight, they have smaller appetites, and can subsist on less food than any variety; indeed, on a good run they will almost get their own living, foraging everywhere, not standing idly about as in the case of the Asiatic breeds, but ever moving.

They are very valuable for an unprotected place, whether it be a wild run, or near a road or a stable yard, both from their activity in avoiding danger themselves and their courage in defending their little ones. They will not bear fattening, but are always full of meat if on a good range, and with suitable food they are plump and meaty from little chickens, while their quick growth and hardihood are two other valuable points not to be lost sight of. They will roost in the trees throughout the winter with impunity, indeed they never do so well, nor show such splendid condition and plumage as when they roost in trees or open sheds, &c., the year round.

Finally, we may say that, although they are unable to stand close confinement for long together, yet there is no fowl so suitable to persons who breed poultry for their own table and have a good run for them. They are the fowls for the country gentleman, for the outlying farmsteads, for the cottager who has a run for them in the green lane or common, for anyone who has a meadow, orchard, or copse, where they can run during the day and forage for themselves, and reward their owner by their healthy appearance, a plentiful supply of eggs and fine condition; their beautiful colours, and their graceful movements, so superior to and more attractive than those of the common breed of fowls. The cocks, unless old, are seldom anything but gallant in searching for food for the hens and calling them to partake of some choice morsel; while, in defence of his *seraglio*, a game cock with full natural spurs has been known to kill a fox; and many a thieving rat has fallen a victim. I have a little Derby hen in my possession, as I write, that killed a weasel in defence of her brood. Seven chickens lay dead, and the hen was wounded by his teeth; but at some little distance lay the weasel, where he had crawled away and died of the wound he had received from the gallant and courageous little hen.

CHAPTER VII.

MANAGEMENT AND MATING OF THE BREEDING STOCK.

IT was the practice of old cockers (and they knew a hundred times more of the art of breeding than we do at present) to put a full-grown stag (or young cock) with six full-blooded sisters of not *less* than two years old, strictly avoiding pullets, unless they were to breed with an old cock of four or five years. Almost all the prominent breeders of the present day object also to breed from pullets, and unless they are *fully* a year old I would certainly not do so, as the chickens are never so strong as from an adult hen. The male bird should be also quite nine or ten months old, and in this case he should be mated to old hens certainly not less than two years of age. The rule of the old breeders in breeding from adult stock on both sides (and this was usually held to produce the strongest birds) was one cock with three hens, and if both birds are over two years old this number should not be exceeded. Personally, I have always noticed the best results from a cockerel with four hens or an old cock with three, or, if not *very* vigorous, two hens.

While we hear a great deal about the degeneracy caused by inbreeding (that is, the breeding together of birds already closely related), it is an undoubted fact that nearly all the best strains of game in England were produced by a system of in-breeding, and old breeders for the pit seldom crossed their stock. I know a case of a gentleman who had some noted fighting birds, and seeing a cock that caught his eye while fighting he bought him, and bred from him and his own hens. The produce proved far worse than either parents, and were almost worthless. An old breeder who crossed his fowls about twenty years ago, assured me he could go on without fresh blood for another twenty years and the stock would show no deterioration. And there is no doubt that where a good many pens are bred of the same blood and a correct account kept of their relationships, the most distant being bred together, it is possible to go on much longer than is generally supposed; provided, of course, that none but the strongest and finest specimens are bred from, and they have good runs, &c. On the other hand, some breeders are always crossing, and our American cousins cross as much as possible, and exclaim against inbreeding as the source of all evils.

In breeding game fowls the male bird has not so much influence as in some other breeds. The limit of weight for cocks was formerly 3 lbs. 6 ozs. to 4 lbs. 8 ozs., when in condition to fight. Birds in ordinary condition would be about 1 lb. heavier, although there were always much larger birds bred, as at present, which were known as turnouts or shakebags, and usually fought in single matches. The cock has most influence in colour of plumage, head and fancy points.

The hen usually gives size and shape, and if spurred, so much the better. Cockerels resemble the father, pullets the mother, as a rule, *to a certain extent*. None but birds of strong constitution and in perfect health should be selected to breed from. An old author on the subject says, "Rottenness requires no cure, but total eradication."

Of course a quiet place, with a good range away from other fowls, is the best place to breed game fowls, and although they may be bred in small runs they never do so well, indeed close confinement for any length of time destroys their constitution, and the hens cease laying. If they have houses to roost in, see that they are *clean, well ventilated*, and *free from draught;* also that they are lighted in some way, as fowls will not resort to a dark house unless compelled. The perches should be round, about the thickness of a man's wrist, fir poles with the bark left on are suitable. If higher than four or five feet from the ground the fowls have not room to fly down, as they have from a tree in the open; consequently they bruise their breasts or feet. The floor should be covered with road sweepings, sand, ashes, or peat moss litter, and the nests for the hens on the ground as secluded as possible, and hidden by a board or other means, that the hens may be quiet and unobserved when they wish to lay. Birds having full liberty will require little besides a few handfuls of corn night and morning, *clean* water, and *cleanliness* in the *house* and *surroundings*, without which no birds will retain their health long. Barley and wheat are the best corn for game; good sound grain, not rubbish; with occasionally oats, peas, or maize; maize, however, is too fattening except in very cold weather, and stops the production of eggs. A little oatmeal or middlings, or, in winter, barley meal mixed with a few boiled potatoes, and a little meat or broth, is good for a morning meal occasionally, but the evening meal should always be good sound corn. Besides this fowls require a supply of green food of some kind; sharp grit, without which they cannot digest what they eat; and some broken oystershell or lime in some form for the formation of eggs. A dust bath is also necessary for them to clean their feathers and free themselves from vermin. Eggs from the breeding pen should be collected at least daily, duly marked at the small end with date and distinguishing mark, and placed in a box in a dry place, imbedded in chaff or bran and turned daily. This is all necessary, and will prevent mistakes in setting stale eggs, &c.

CHAPTER VIII.

SITTING, HATCHING, REARING, &c.

IF the eggs are to be set under the hen that has laid them, she should be in a quiet place, and undisturbed by other hens. If, however, as is often the case, a strange hen is obtained, a small box having no bottom, or a board bottom if preferred, should be procured; it should be large enough for the hen to have room to turn without crumpling her tail feathers, as unless she is comfortable she will probably not sit steadily. A quiet place is desirable for hatching, a turf two inches thick, the size of the bottom of the nest, may be used for a foundation, or put in the nest a good shovelful of earth. Beat this down well with the hand to resemble the shape of a saucer, the centre being the lowest part. Get some dry fern or oatstraw, well rub with the hand to make it soft and lie close, and make a nice nest; place two or three nest eggs in it, and in the evening put your hen quietly on and shut her in, either by a sack over the front, or other means. Next day she should be taken off to feed, and if she does not go back in about twenty minutes she should be caught as quietly as possible and replaced and covered in as before; the second morning she will generally return of her own accord. Sitting hens should be fed once a day, in the morning, with corn and water; they can be all taken off together, and at the end of twenty minutes those that have not returned may be gently driven towards the nests, when they will generally go on. They can then be shut in, and the dung, &c., swept up and the place kept clean. Should an egg get broken in the nest the others may be put into a bucket of warm water at about 70° Fahr., and gently washed clean and put back under the hen, otherwise they will all stick together and be constantly breaking, or sticking to the feathers of the hen. On the twenty-first day the chickens usually make their appearance; some latitude should, however, be allowed, as sometimes they come out a day or so earlier, and are sometimes a day or two behind time. Therefore if a valuable brood is expected, and only a few are hatched out, the remainder should be put under another hen, as the first, finding the chickens moving beneath her, will raise herself up and allow the remaining eggs to get chilled.

The chickens hatched out should be left undisturbed for twenty-four hours, they may then be taken and placed under a coop with the hen in a dry place. If the coop has a board bottom it should be covered two inches deep with dry earth to prevent cramp in the chickens. Chickens can stand cold, but *none* can withstand *damp*. The first food should be finely chopped hard boiled egg and stale bread crumbs, moistened, if desired, with a little new milk. They should have food every two hours, and if early in the year, should be fed the last thing by lamplight. At the end of a fortnight every three hours will be sufficient, and after the first month four or five times a day. After the first week coarse oatmeal, boiled or raw, mixed with fresh milk, a little canary or hemp seed, and animal food occasionally, grit and green food, once a week a little boiled rice may be given, Spratt's meal and Crissel, and dog biscuit broken up small and soaked, are also good food. Change is the great thing, and to give them just as much as they will clear up and no more. From the first few days corn may be gradually begun, such as small broken wheat, &c.; this should be given at night or the last feed. If water is given it must be kept out of the sun and changed frequently, otherwise it will produce diarrhœa. Many breeders give no water until the chicks are two or three months old, but let them have milk to drink twice a day, taking it away as soon as they have had sufficient. Frequently at this age the whole brood will commence fighting and one will attain the mastery, which, if not interfered with, he will keep until they are five or six months old. In this, however, there seems no rule, some will fight so that one or two of the brood may be spoilt or killed at seven or eight weeks old; at other times they will remain peaceably together until as many months. At the age of ten or eleven weeks they should be allowed to roost. The hen will go to roost and they will crowd round on each side. If they do not roost in the trees they must have a perch suitable to their size, as if too large some will become duck-footed, that is, the hind toe brought round sideways, generally caused by the perch being too big for them to grasp. The breeder should occasionally just feel their breast bones as they are on the perch.

If there is the slightest deviation from the straight line, or any indentation the bird should be put in a hamper for a week or so at night, and not allowed to roost, the bone will then probably have recovered its form, if not too far gone at first. The large breeds of game are more subject to this than the smaller birds. A good old cock running with cockerels will generally prevent them from fighting, in fact will not allow them to do so, but if at all savage will strike and permanently injure some of them. It is well, therefore, if running with cock chickens, to remove his spurs.

At the age of about six months, or when the comb and wattles are fully grown, the cockerels should be dubbed. An attendant should hold the bird, the operator, with a pair of very sharp scissors, should take off the wattle close, but not pull it at all. Then taking hold of the lower beak with one finger in the bird's mouth, and the thumb at the back of his head, he should, with his right hand, commence at the back of the comb close to the head, and with one cut take the comb fairly close off to the beak; the ear-lobes may then be just trimmed off, and the bird tossed up, when, if a few grains of corn are thrown down he will generally commence to eat, or will eat his own comb if he has the chance. There is *very* little pain attending this operation, if *properly done*, and it takes less than a minute. The beginner should get some old hand to show him the process, as he will learn better by seeing than by any written instructions. Some take off the comb *quite* close to the head, others leave a little; this is a matter of taste. I dislike myself to see them cut too close. After dubbing the young cocks will commence to fight and must be divided, or sent out on separate walks.

Some breeders separate cockerels and pullets at an early age, and a lot of cockerels will run peaceably together for some time provided all sight of a hen or pullet is kept from them; the sight of one would be the signal for a battle royal. Opinions differ as to whether finer birds are obtained by the separation of the sexes; it is generally thought that this is the case. One caution is necessary,—it is impossible to breed fine fowls, or to keep healthy stock if there is any over-crowding either on the ground or in the houses. It is better to have twenty healthy birds than a hundred diseased, and all sorts of complaints will appear if too many are kept. The smaller the number the better they thrive, *always*.

CHAPTER IX.

EXHIBITING.

WHEN the amateur thinks he has a bird or birds good enough to make a creditable appearance at a show, he should take care that the show where he enters them is one where a competent judge is officiating. The birds will probably require no preparation, if on a good run, until the day before the show, when the legs, feet, beak and face should be washed; the legs and feet well scrubbed clean with soap and water, a piece of flannel with "just a suspicion of vaseline" or olive oil, rubbed over and wiped off again with a clean handkerchief. The face may be just touched with vinegar and a little oil, and wiped clean, which will brighten the red, and the bird put into a well lined hamper with plenty of straw at the bottom, and dispatched by a train that will take them to the show as quickly as possible.

In bad weather they sometimes require shutting in a clean room or shed with clean straw for them to scratch in for a few days previous to the show. Oatmeal and Indian corn boiled together to a jelly, with a little linseed, a feed of split peas at night, and grit, green food and clean water with a little sulphate of iron in it, will keep them in good condition and spirit. On returning from the show a feed of soft food should be given, after which they can be returned to their usual runs. It is allowable to remove the feathers standing up on each side of the cock's comb; further than that no trimming is permissible.

CHAPTER X.

DISEASES.

THERE is no need to give here a full description of the treatment of diseases. Several works are published on the subject, within the reach of all. We will just mention a few of the most common ailments.

Moulting is not a disease, but sometimes requires treatment A little extra nourishing food, a feed of hemp-seed, and once or twice a week a little sulphur in the soft food may be given, and if the tail feathers do not grow properly a little vaseline should be applied to the roots. If any vermin are present powdered sulphur or insect powder dusted into the feathers will destroy them.

Bumble-Foot is caused by high perches or hard floors, the birds coming down heavily on their feet, or treading on some sharp substance, or bruising the sole of the foot. In the early stages painting with tincture of arnica is good; if advanced and matter formed two cuts should be made crosswise, the matter squeezed out and the foot fomented with hot water, and a dressing put on consisting of one part carbolic acid and ten parts olive oil. If this does not effect a cure blue stone or caustic must be applied or the foot poulticed with linseed meal. The bird should be kept on a bed of soft straw or moss, and not allowed to roost until well.

Roup.—This is the worst disease to which poultry are subject; any bird showing symptoms of it should at once be removed from the rest, and completely isolated. The symptoms of roup are an offensive smell arising from the nose and mouth, frothy bubbles, and sometimes pus at the corner of the eyes, round which the head is swollen, and sometimes the bird is unable to see. A solution of sulphate of copper to wash eyes, nostrils, and mouth is recommended, as is also a weak solution of carbolic acid. Pills consisting of half a grain of sulphate of copper, one third of a grain of hydrastin, oil of copaiba three drops, powdered charcoal, and ginger are recommended to be given twice daily. Another remedy is to syringe the mouth and nostrils with water in which sufficient permanganate of potash to give a rich colour is dissolved. Another—equal parts pulverised alum, acetic acid, and sugar of lead, to be used in the same manner.

Cold.—Sometimes comes on suddenly, with running at the nostrils and swelling round the eyes, resembling roup, into which it will develop if not checked. If the face is swollen foment with hot water, wash out nostrils and roof of the mouth with very weak solution of carbolic acid in rain water, give three or four drops camphorated oil, and a teaspoonful of glycerine for two or three nights and some finely chopped green rue made into pills with a little butter daily; give soft food, and keep warm and free from draughts.

Indigestion.—Give the bird a teaspoonful and a half of olive oil, and feed entirely on soft warm food for a while until recovered.

Canker.—Paint daily with carbolic acid one part, glycerine ten parts, first scraping off the discoloured portions.

Scaly leg.—Thoroughly cleanse the affected parts with soap and water and a hard brush, and thoroughly work into the scales an ointment made of sulphur and soft paraffin.

Bruises and wounds should be well bathed with hot water, or warm water and milk, and wounds dressed with carbolic acid one part, and olive oil fifty parts.

Gapes are confined to chickens; subjecting them to the fumes of carbolic acid will effect a cure. Chickens should have frequent attention to discover if they are infested by lice, which in the warm weather kill many, as they become thin and have no strength.

Flowers of sulphur dusted well into the feathers, or an ointment of sulphur, lard, and paraffin, rubbed on the head and under the belly will destroy them.

INDEX.

	PAGE
Agreement for Cock Match	48
Battle Royal	49
Birchens	33
Blacks	23, 27, 36
Black-breasted Reds	20, 28, 29, 30, 31
Black-breasted Silver Duckwings	23
Blue-reds	21
Brass Wings	24
Breeding	46, 57
Breeds and Colours	26
Bright-reds	20
Brown-reds	21, 31
Bruises and Wounds	65
Bumblefoot	64
Canker	65
Chickens	61
Cock Fighting in England	13
Cocking	43
Cock Pits	14
Cold	65
Cuckoos	37
Diseases	64
Dubbing	62
Duckwings	22, 32
Duns	27, 35
Early Records of Cock Fighting	10
Exhibiting	63
Feeders	46
Feeding	47
Game in Confinement	56
Gapes	65
Ginger-reds	31
Greeks	10
Greys	27, 32
Hatching	60
Heeling	45
Hennies	27, 37
Indian Cock Fighters	10
Indian Game	27
Indigestion	65
King Charles II. and Cock Fighting	13
Laws of Fighting	51
List of Colours	25
Lord Derby's Reds	28
Mains	49
Management	57
Modern Game	16
Moulting	64
Muffs	27, 41
Old English Game in the Show Pen	16
Origin and History	16
Piles	22, 27, 34, 35
Points of Game Fowls	19
Reds	26
Romans	10
Roup	64
Rules for Matching and Fighting	49
Scale of Points	24
Scaly Leg	65
Sitters and Mothers, Old English Game as	55
Sitting	60
Spangles	24
Sport of Cocking	43
Spurs	44
Standard	19
Tassels	27, 42
Useful Properties	55
Welsh Main	49
Whites	23, 27, 36
Wild Varieties	9
Yellow Duckwings	22, 33

Advertisements.

CHAMBERLIN'S CANADIAN POULTRY MEAL,

THE CHEAPEST FOOD IN THE WORLD FOR

TURKEYS, GEESE, DUCKS, AND CHICKENS.

Received the ONLY AWARD given by the International Jury, Paris Exhibition, for Poultry Food.

Bronze Medal and Diploma, Mannheim, 1880. Silver Medal, Antwerp Exhibition, 1885.

TURKEYS, Geese, Ducks, and Chickens fed on this Meal thrive wonderfully, are kept free from Disease, and will lay nearly double the number of eggs. Fowls commence laying about seven months old. Turkeys, so difficult to breed, are reared with the greatest success upon this Meal, with scarcely a loss of three per cent. If shut up they will fatten in a very short time, and the colour and delicacy of the meat is surprising. It is invaluable in cold and exposed situations, and success at Poultry Shows is also assured by the use of this Meal.

Numerous Testimonials on Application.

Price 20s. per cwt., including 12 packets of Aromatic Compound and bag. Packed also in Three-pound Packets, 6d. each.

KALYDÉ,

An Infallible Cure for Gapes in Pheasants and Chickens.
2s. per Tin; Post Free, 2s. 6d.

ROUP PILLS,

For Poultry, Pigeons, and Cage Birds.

CONDITION PILLS,

For Poultry, Pigeons, and Cage Birds.

Above Preparations, 1s. per Bottle; Post Free, 1s. 2d.

JAMES CHAMBERLIN & SMITH

(Late JAMES CHAMBERLIN),

Game, Poultry, and Dog Food Warehouse,

POST OFFICE STREET, NORWICH.

CHAS. FRAZER'S Exors., Manufacturers, Norwich.

All kinds of Poultry Houses made to order.

No. 72.

New Combination Fowls' House. Cash Price complete, £7 10s. Carriage Paid.

No. 74. Portable Fowl House & Run.

CASH PRICES. Carriage Paid.
10ft. by 5ft. (including run) ...£4 10 0
12ft. by 6ft.   ,,   ,,   ... 5 10 0
15ft. by 6ft.   ,,   ,,   ... 7 10 0

No. 74a. Portable Duck House.

Size—6ft. long by 2ft. 6in. wide, 3ft. high.
Cash Price, Carriage Paid ...£2 7 6

No. 75. Fattening Pen.

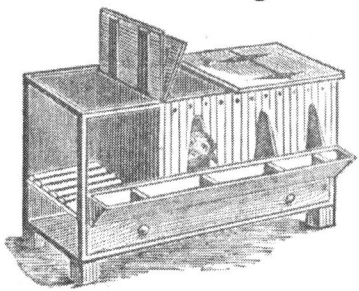

CASH PRICE.
For Four Fowls £1 8 6

No. 73. Chicken Coop.

No. 76. Folding Coop.

CASH PRICES.
One Coop, 7/-. 6 Coops, 40/-. 12 Coops, £4.
Wire Netting Bottoms to prevent rats burrowing underneath, 9d. extra.

All Orders of 40s. value and upwards sent Carriage Paid to any Goods Station in England and Wales. Complete Catalogue of Conservatories, Green-houses, Garden Frames, Propagators, Hand-lights, Poultry Appliances, Pigeon Cotes, Dog Kennels, Garden Seats, Chairs, &c., Post Free for Six Stamps.

No. 71. Hatching and Nesting Box.

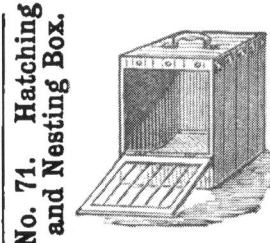

CASH PRICES.
One Nest, 4/-.
Six Nests, 22/6.
Twelve Nests, 44/-.

Advertisements.

𝔊𝔞𝔪𝔢 𝔉𝔬𝔴𝔩𝔰 𝔈𝔵𝔱𝔯𝔞𝔬𝔯𝔡𝔦𝔫𝔞𝔯𝔶!

MR. JOHN BROUGH,
22, LONDON ROAD,
CARLISLE CUMBERLAND,
Breeder of Thorough-bred Game Fowls.

OFFERS FOR SALE.—Cocks, Cockerels, Hens and Pullets, of the following colours:—Blacks; Brassy-winged; Black, B. Reds; Brown, B. Reds; Blue, B. Blue Reds; Duns, Spangles; Cuckoos; &c., &c. These have been carefully bred from the old strains, which my father fought so successfully in the London, Liverpool, Manchester, Newcastle and Newmarket Cock-pits, before they were suppressed, and latterly have been most suecessful in the Old English Game Classes at all the leading Shows, including 3 Specials, 5 Firsts, Wigton; First, Birmingham; and three years in succession in Cockerel and Pullet Classes at the ROYAL SHOWS. Reference as to their game quality and purity of breed, to *GAMECOCK*, of the "Fanciers' Gazette," &c.

Advertisements.

Important to Poultry Keepers and Game Raisers.

BROWN'S AROMATIC COMPOUND
For Poultry, Game, &c.

(Is found of immense benefit by all Keepers of Birds who use it.)

The best Article ever sold to secure Eggs in cold weather; Assists the process of Moulting; Should be given to all Birds put up for Fattening.

Has stood the test of nearly a quarter of a century. Testimonials from all the leading breeders. The cost is merely nominal; a 10/- tin sufficient for 1,000 head for a week; 1s. per week for 100.

Sold in Canisters:—No. 1, 1/3 each; No. 2, 3/- each; No. 3, 5/6 each; No. 4, 10/- each; No. 5, 20/- each; the larger sizes being greatly cheaper than the smaller ones. All carriage paid.

CAPSULES OF GUARANTEED QUALITY.

The purest and most effective Medicine known—Cod-liver Oil, Cod-liver Oil with Quinine, Castor Oil, Charcoal, Astringent, Copaiba, Turpentine, Areca Nut, Oil of Male Fern, and Santonin. Made in five sizes.

FULL LISTS ON APPLICATION.

1/- per Box, post free, 1/1½; 6 Boxes 5/-, post free

CANARYPER, for Colouring Cage-birds, &c., &c., used successfully for 10 years at home and abroad.

Sold in Tins:—¼-lb., 1/3; ½-lb., 2/6, post free; 1-lb., 4/6; 4-lb. sent carriage paid to any address for 15/-.

E. T. BROWN & SON,
31, Dean Street, Newcastle-on-Tyne, England.

Wholesale Agents—LONDON: BARCLAY & SONS, Ltd., Farringdon St., E.C.; SANGER & SONS, Oxford Street, W.
CROYDON: JOHN WALTON, St. James's Road.

Advertisements.

John Wilson Simpson,

ABBEY TOWN,

SILLOTH, CUMBERLAND,

The most Successful Breeder and Exhibitor of Old English Fighting Game.

Scores of Cocks have been supplied by me for the pit during the last few years, and only one single Cock has ever lost a battle that I have been informed of.

Cocks and Hens sold by me have won Specials, Firsts, Cups and other Prizes at many Shows, including Royal Shows, 1889-90-91; also Birmingham and other leading Shows.

Good Birds, both sexes and all ages, always on hand, including many old noted Strains.

Advertisements.

GEO. MATTHEWS,
VELINDRE, KNIGHTON, RADNOR,
Breeder of Pure Old English Game.

BLACK REDS AND PILES have been carefully bred for years from the best blood in England.

J. D. TOOGOOD PARSONS, Jun., Esq.,
ASHURST PLACE, LANGTON, TUNBRIDGE WELLS.

BREEDER OF
Black Red Old English Game (White-legged),
Lord Derby and other noted Strains.

THOMAS ROPER,
WETHERAL, CUMBERLAND,
Breeder & Exhibitor of Old English Game

Birds for Stock, Exhibition, or Pit purposes always on hand.

EGGS IN SEASON.

ALSO STEEL SPURS OF FINEST QUALITY.

Important Notice to Keepers of Poultry, Pigeons & Cage-Birds.

THE EXCELSIOR POULTRY PILLS

Having proved the best Pill ever prepared for Poultry, should be tried by all Poultry Keepers. They have cured thousands of Poultry and Birds of all kinds of the following complaints: Cold, Roup or Croup, Chip, Crampy Spasm, Scour or Flux, Scale or Pip, Swollen Eyes and Head, Turn or Giddiness, Canker, Wasting away, Wind Eggs, Slow Moulting, &c. They are a wonderful Pill. By using **Excelsior Poultry Pills** you gain healthy birds, plenty of eggs, and strong chicks. If you cannot get them from your Chemist or Corn-dealer, I send three 2d. boxes for 7d., or one case of eighteen for 3s.; two 6d. boxes for 1s. 1½d., or one case of twenty-four for 11s., carriage paid. **Excelsior Poultry Liniment**, for Sprains, Swellings, Bumble Foot, Cramp, Gapes, Leg Weakness, Rheumatism, Canker, Scaly Legs, Wounds, Sores, Spotted, Ripped, Cracked, and Frostbitten Combs and Toes. The best Liniment in the World for all kinds of Pets. One Bottle 1s. 2d., Four Bottles 4s., post free.—W. H. LAKIN, 16, New Bond Street, Leicester.

Advertisements.

LEGHORNS.

L. C. VERREY,

Breeder and Exhibitor of

WHITES, BROWNS ❖

AND THE NEW

❖ VARIETY BUFFS.

BIRDS FOR DISPOSAL

At Reasonable Prices for Exhibition or Stock.

Many Birds from this Yard exported to the Continent and the Colonies, where they and their progeny have proved successful in the Show-Pen.

ADDRESS—

Oak Lawn, Leatherhead,
SURREY.

Advertisements.

R. STACEY,
Heatherlands, Tilford, Farnham,
SURREY,

CHAMPION STRAIN OF PLYMOUTH ROCKS.

Birds of this famous established medium colour Strain have been wonderfully successful in all parts of the world; also 1st, 2nd and 3rd Prizes at the Crystal Palace, 1887; 1st & Cup, 1888; Two 1sts & Two Cups, 1889; Silver Medal, 2nd and 3rd, 1890.

BIRDS OF ALL AGES FOR DISPOSAL.

Our Works are in the Country, *close* to a Railway in the Midland Counties, within easy reach of the principal timber ports, and by the aid of powerful Sawing, Planing, and other Machinery we can defy any Maker on Earth to compete with us for Lowness of Price and High Quality.

		Height.	Length.	Width.	Price.	Floor ext.
No. 5..	For 10 to 14 Fowls..	5ft.	3ft.	4ft.	15/-	3/-
,, 1..	,, 12 to 20 ,,	6ft.	4ft.	4ft.	21/-	4/-
,, 2..	,, 20 to 30 ,,	6ft.	6ft.	4ft.	27/-	6/-
,, 3..	,, 30 to 40 ,,	6½ft.	6ft.	5ft.	30/-	8/-
,, 4..	,, 50 to 60 ,,	7ft.	8ft.	6ft.	40/-	10/-

Testimonial.

Laureston Vicarage, Tavistock, Feb. 5th, 1891.

DEAR SIR,—I am much pleased with the Rabbitry (consisting of 24 hutches), and Cavy House you have sent me, and think them well worth the money I have paid for them. You will hear from me again when I want anything of the kind.—Yours faithfully,
J. H. W.

E. C. WALTON'S FOWL HOUSES

Are complete with window and slide, nest boxes, perches, ladder, door for attendant, with lock and key, ventilator, and a thoroughly watertight roof. If they are not as represented, are returnable. They are made in sections, to pack flat for travelling, and can be put together in a few minutes.

E. C. WALTON, MUSKHAM, NEWARK.

CALWAY'S HOUSES & APPLIANCES ARE BUILT ON THE Most approved Principles, And are universally received as the best in the market. They are always sent on approval.

Quotations for any kind of House or Appliance will be given by return of Post on receipt of requirements.

Advertisements.

The Fanciers' Library

Andalusian Fowl. By L. C. VERREY, 1s.

Leghorn Fowl. By L. C. VERREY, 1s.

Plymouth Rock. By J. WALLACE, 1s.

The Old English Game Fowl.
By HERBERT ATKINSON, 1s.

Wyandotte Fowl.
By J. PENFOLD FIELD, 1s.

Either of the above, post free, 1s. 2d.

French Breeds of Poultry.
By L. C. VERREY, 1s. 6d., post free, 1s. 8d.

Successful Pigeon Culture.
By RICHARD WOODS, 2s., post free 2s. 2d.;
cloth, 3s. post free.

Rational Breeding.
By E. MILLAIS, 2s., post free, 2s. 3d.

The British Canary.
By C. A. HOUSE, post free, 7d., cloth, 1s. 8d.

Poultry Keeping as an Industry for Farmers and Cottagers.
By EDWARD BROWN, F.L.S., 6s., post free.

The FANCIERS' GAZETTE, Ltd.,
54 to 57, Imperial Buildings,
LUDGATE CIRCUS, LONDON, E.C.

Advertisements.

Every Friday.]　　Established 1874.　　[*One Penny.*

The Fanciers' Gazette

A Weekly Journal devoted to Dogs, Poultry, Pigeons, Rabbits, Cats, Cavies, and Cage-Birds.

Articles by the Leading Authorities;
　　Letters and Notes on Current Subjects;
Prompt and Reliable Reports;
　　Queries answered by Specialists;
　　　　Illustrations by best Artists.

ONLY PENNY JOURNAL DEALING WITH ALL THE FANCIES.

TERMS OF SUBSCRIPTION
(Payable in advance).

	s.	d.
One Copy, per post	0	1½
Six months, do.	3	3
Twelve months, do.	6	6
Do., India, do.	10	10
Do., other Foreign Countries	8	8

RATE FOR ADVERTISEMENTS.

	s.	d.
Per line (containing on the average six words)	0	6
Shows, and Stud Dogs :—		
1 to 3 insertions, per line	0	6
4 to 6　,,　　,,	0	5
7 to 12　,,　　,,	0	4
13 and upwards　,,	0	3
Auction Sales, per line	0	9

Discount allowed on series, prepaid.

Per inch in column (12 lines to the inch)	6	0

Contracts made for series.

SPECIAL ADVERTISEMENTS

For Sales, Purchases, or Exchanges, One Penny for every Three (or part of Three) Words.

No advertisement is received for a less sum than Fourpence. Addresses charged. Remittances must accompany order.

Address Cheques, Post Office Orders, and Communications to—

The "FANCIERS' GAZETTE," Limited,
54 to 57, Imperial Buildings,
　　　　Ludgate Circus,
　　　　　　London, E.C.

20 PER CENT. SAVED
BY USING THE
CEREAL MEAL.

The most Complete and Successful FOOD of the day in rearing

FOWLS,
PHEASANTS,
TURKEYS
and DUCKS.

Ensures Easy Rearing, Quick Maturity and Vigorous Stock in all Varieties.

Invaluable for the Production of FERTILE EGGS.
Stands UNRIVALLED for Completeness and Economy.

Specially adapted for Preserving Exhibition Stock in Good Condition.

The result of 14 years of Practical Study in the Manufacture; and attested by all the leading authorities as the best in the Market.

Can be had only through duly appointed Agents or direct from the Manufacturer:—

THOS. LAMBERT,
POULTRY & GAME'S FOOD MANUFACTURER,
BOURNE MILLS, HADLOW, KENT.

Ask for the CEREAL and see that each Bag is Stamped with above Trade Mark.

J. HURST, Esq., winner of two Challenge Cups and Cup for best Leghorn at the late Palace Show writes:—"I have still the good opinion which I always had of your Cereal Meal, and which I think is the best Meal in the Market."

Printed in Great Britain
by Amazon